Generis
PUBLISHING

AF589263

ASSESSMENT OF THE EFFICIENCY OF CRUSHED CERAMICS IN ADSORBING METHYL ORANGE DYE FROM WASTEWATER

Timilehin Francis Olaleye

Title: **ASSESSMENT OF THE EFFICIENCY OF CRUSHED CERAMICS IN ADSORBING METHYL ORANGE DYE FROM WASTEWATER**

ISBN: 979-8-89248-401-5

Author: Timilehin Francis Olaleye

Cover image: https://unsplash.com/

Publisher: Generis Publishing
Online orders: www.generis-publishing.com
Contact email: info@generis-publishing.com

DEDICATION

This project is dedicated to the glory of God in Heaven and the benefit of mankind on earth.

ACKNOWLEDGMENT

All glory and gratitude to GOD, who has been my help, strength and source of wisdom throughout the course of this project work.

I appreciate the encouragement and unflinching support and love of my parent Chief and Mrs. Olaleye, their moral, financial support since I started my educational pursuit.

I also want to use this medium to appreciate all chemistry laboratory technologists, for their inputs during my bench work which has contributed to the success of this project.

Finally to my colleagues and friends, I appreciate your support and advices, God bless you all.

ABSTRACT

Objective: This study investigates and assesses the potential of crushed, pulverized ceramics as a low-cost adsorbent for the removal of methyl orange from wastewater. The presence of heavy metals and dyes in water bodies are deadly to the living organisms inside water, in which these pollutants are bio-accumulated and biomagnified in the environment. Due to the health effects of these pollutants, it is, therefore, necessary to treat metal and dye-contaminated wastewater prior to its discharge into the environment in order to comply with the stringent environmental regulations and also safeguard the present and future generations.

Methods: The influence of pH, contact time, initial metal concentration, adsorbent dosage and temperature were studied in batch experiments at room temperature and were measured using UV-VIS Spectrophotometer at wavelength 464nm. Fourier Transform Infrared (FTIR) techniques was employed as an instrument for characterization of the adsorbent before and after adsorption and the data were collected and interpreted using Microsoft Excel, 2016.

Results: Maximum sorption for methyl orange was found to be at pH 2. The adsorption was rapid at the first 90 min of contact, with uptake of more than 90%, and equilibrium was achieved in 60 min of agitation. Langmuir, Freundlich and BET's isotherm models were applied to describe the adsorption of methyl orange dye. Fourier Transform Infrared (FTIR) spectra of ceramics powder revealed that OH, C-H, C=C, C-O stretching were responsible for the adsorption. However, the effects of different experimental parameters that influenced the efficiencies of the adsorbent have been evaluated and optimized.

Conclusion: The investigation revealed that the adsorption capacity of the powdered ceramics on the removal of methyl orange dye is high enough compared to observed values in literatures. Freundlich's model fitted the equilibrium data better, while the pseudo-second-order kinetic model was the most fitting from the kinetic data obtained for the adsorption of methyl orange dye.

Keywords: Ceramics, Methyl orange, Dye, Pollutants, Adsorbent, Adsorption

TABLE OF CONTENTS

LIST OF FIGURES

LIST OF TABLES

CHAPTER ONE.

INTRODUCTION

1.1 Background of Study

Proper health cannot be guaranteed, due to pollution everywhere. Environmental pollution is the activities that lead to the contamination of the physical and biological components of the earth in which normal environmental processes are adversely affected. Environmental pollution has been a great disaster to the world which has declined the population rate of people. It has majorly been caused by anthropogenic activities which resulted to about six kinds, they are; air, land, noise, thermal, visual and water pollution.

Water is the major natural resources that play the central role in mediating global scale ecosystem processes, linking lithosphere, atmosphere and biosphere by moving substances between them and enabling chemical reactions to occur. Water is essential to both natural ecosystem and to the human development, but now, it has been a threat to the health of the environment due to pollution (Aslam *et al.*, 2004).

Water pollution is majorly caused by industries in which many industries do not care about the environment before discharging their wastes into the water bodies. Dye producing industries such as dyestuff, textile, paper and plastic use dye to improve the colour of their products and consume substantial volumes of water. These industries generate large amount of coloured wastewater. The presence of small amount of dye in water makes the water to be undesirable and dyes are toxic, carcinogenic and pose a serious threat to living organisms (Issabayeva *et al.*, 2010). Living organisms need water for biochemical processes; therefore, there is a need to remove dyes before effluent is discharged into receiving water bodies.

1.2 Problem Statement

Environmental problems arise mainly from inefficient removal of dyes from effluent receiving bodies. Dye manufacturing industry, textile industry and user of fabric are major contributors to discharging of dye effluent into water bodies. It is estimated that dye manufacturing industry discharged 2% of dye produced annually in effluent from manufacturing operations. While in the textile industry, 10-15% of the dye is lost during the dyeing process and released with the effluent (Zollinger, 1991). Dyes are also found useful in other allied-industries involved in the manufacturing of laboratory equipment, paper, plastic, fertilizer, detergent, leather, dyestuff, cosmetics and pharmaceuticals. There are more than 10,000 commercially available dyes, in which most of the dyes produced have side effects on the living organisms (Pinherio *et al.*, 2004).

Synthetic dyes are more stable and difficult to biodegrade because of their complex aromatic molecular structures. On exposure of living organisms to these dyes will surely lead to disorders of their bodies, because they are toxic and carcinogenic (Kamaljit Singh *et al.*, 2011). Organic dyes are also hazardous and affect the aquatic life and the food chain. Azo dyes are more carcinogenic to human health.

Physical, biological and chemical methods and technologies are commonly applied in removing dyes from aqueous solution. However, the application of some of these methods may be impractical due to economic constraints, insufficient to meet strict regulatory requirements and they may generate hazardous products which are difficult to treat (Reddad *et al.*, 2002).

Among of all these methods, adsorption is the most effective and non-destructive techniques that is widely used for the removal of dyes from aqueous solution due to its initial cost, simplicity of design, use of operation and insensitivity to toxic substances (Uddin *et al.*, 2009). Adsorption is extensively used in industry for separation and purification of wastewater. Using of activated carbon as the adsorbent has been shown interested in due to its good capacity for adsorption of dyes but its usage has been limited because of its high cost and about 10-15% loss during the regeneration (Bhattacharyya and Sharma, 2003).

Presently, researchers are interested in using alternative adsorbents for removing dyes; with the key characteristics required being due to reusability, low operating cost, improved selectivity for dyes of interest, removal of dyes from effluent irrespective of toxicity, possibility of reuse, short operation time, and no production of secondary compounds which might be toxic. Some of the materials that have been studied for the removal of dyes include various biosorbents- like mango peel, sugarcane bagasse, corn cob, coconut shell, clay, silica etc.

This present research work studies the use of crushed ceramics for the removal of dyes, specifically, methyl orange from wastewater.

1.3 Justification of the Study

The presence of heavy metals and dyes in water bodies are deadly to the living organisms inside water, in which these pollutants are bio-accumulated and biomagnified in the environment. Due to the health effects of these pollutants, it is therefore necessary to treat metal/dye contaminated wastewater prior to its discharge into the environment in order to comply with the stringent environmental regulations and also safeguard the present and future generations (UNEP, 2008). The use of alternative adsorbents has provided the best solution to eradicate or reduce the amount of heavy metals and dyes in the wastewater to permissible limits laid by international bodies (WHO and USEPA). It is safer for use than other conventional methods which produced toxic sludge; while it also helps to save foreign exchange that would have been used to import technology and materials for pollution control.

1.4 Aim and Objectives

The aim of this research was to evaluate the efficiency of crushed ceramics in adsorbing methyl orange dye from aqueous solutions, while determining the optimum operating parameters for the process. The specific objectives are:

- To prepare an adsorbent out of broken ceramics, and characterize the adsorbent.
- To study the effect of pH, contact time, adsorbent dosage, and initial concentration on the adsorbent's capacity to remove methyl orange dye from aqueous solutions.
- To investigate the equilibrium removal conditions for methyl orange in terms of adsorption isotherms using Langmuir, Freundlich, Brunauer-Emmett-Teller (BET) isotherm model to fit experimental data.
- To investigate the kinetics of methyl orange removal from prepared solutions by fitting with pseudo first order reaction, pseudo second order reaction and Intra particle diffusion model.

CHAPTER TWO.

LITERATURE REVIEW

2.1 Dyes in the environment

Dyes can be defined as substances that are intensely coloured, soluble, giving transparent solutions, and can be applied in solution or dispersion to fabrics or surfaces to give coloured materials considerable fastness.

Dyes can be classified as natural and synthetic dyes depending on their origin. Natural dyes are gotten from the nature i.e. vegetable matter, insects and minerals. While synthetic dyes are gotten during synthesis which may be water soluble, water insoluble and in-situ colour formation. Synthetic dyes are extensively used in many fields of up-to-date technology, e.g., in various branches of the textile industry (Gupta *et al.*, 1992; Shukla and Gupta, 1992 and SokolowskaGajda *et al.*, 1996), of the leather tanning industry (Tünay *et al.*, 1999 and Kabadasil *et al.*, 1999) in paper production (Ivanov*et al.*, 1996), in food technology (Bhat and Mathur, 1998 and Slampova *et al.*, 2001), in agricultural research (Cook and Linden, 1997 and Kross *et al.*, 1996), in photo-electrochemical cells (Wrobel *et al.*, 2001).

Different types of synthetic dyes are mostly used by textile industries to improve the colour of fibres (Table 2.1).

These pollutants entered into the environment by means of discharging of effluent containing small/large amount of the pollutants. Once, they are into the environment, they become persistence to the environment due to their complex aromatic molecular structure.

Table 2. 1. The estimation of the degree of fixation for different dye-fibre combination and loss to effluent

Synthetic dyes	Fibre	Degree of fixation (%)	Loss of effluent (%)
Acid	Polyamide	89-95	5-20
Basic	Acrylic	95-100	0-5
Direct	Cellulose	70-95	5-30
Disperse	Polyester	90-100	0-10
Metal-complex	Wool	90-98	2-10
Reactive	Cellulose	50-90	10-50
Sulphur	Cellulose	60-90	10-40
Vat	Cellulose	80-95	5-20

Source: Ramachandra TV, Ahalya N & Kanamadi RD *et al.*, 2003

2.1.1 Toxicological aspects of dyes

Dyes usually have a synthetic and complex aromatic molecular structure, which makes them to be stable and difficult to biodegrade. Each dye has particular effects they cause to the living organism that expose to them. For examples, Azo-dyes undergo biological/photochemical degradation when they reach the aquatic environment. The degradation products are more harmful and persistent than the parent compound (Sanchez *et al.*, 2006).

The presence of dye in water affects the photosynthetic activity in aquatic life by reducing the light penetration which serves as source of energy to most aquatic organisms. It causes depletion of dissolved oxygen (DO), which implies that there will be increase in the chemical oxygen demand (COD) and biological oxygen demand (BOD) that refers to the amount of oxygen required to completely oxidize the organic substances in water chemically and biologically respectively. The depletion of dissolved oxygen (DO) makes the anaerobic bacteria to take over which leads to formation of CH_4, H_2S and NH_3 (Khopkar, 2004). These products formed are toxic to aquatic organisms and human that consumes the fish that accumulated it.

2.1.2 Treatment technology for removal of dyes in wastewater

Preventing dyestuff effluent from entering the environment is paramount because of its toxicity and effects they caused to living organisms. The elimination of the dyestuff content from the effluent is done mostly by conventional methods, which are;

2.1.2.1 Physicochemical methods: These are the methods that joined both physical and chemical means of separating dyes from the effluent. They include; ion flotation, solvent sublation and electrocoagulation.

2.1.2.2 Photo catalysis and Oxidation methods: Commercial dyes have high resistance to photo degradation, which has led to the development of catalysts and oxidizing agents for decolourizing of dye wastewater. The catalysts and oxidizing agents such as Hydrogen peroxide with Iron (III) and ozone were reported effective in degradation of the dye intermediate anthraquinone-2-sulphonic acid sodium salt (Kiwi *et al.*, 1993) and decolouration of Orange II. Oxalate, formate and benzene sulphonate ions were the most important decomposition products (Tang and An, 1995a and Tang and An, 1995b) respectively.

2.2 Methyl Orange

Methyl Orange (MO) is selected as the adsorbate in this adsorption studies representing anionic organic pollutants as they are easy to be analyzed by UV-Vis spectrophotometry. Methyl Orange (Sodium 4-[(4-dimethylamino) phenyldiazenyl] benzenesulfonate) is a water soluble, anionic, azo dye with the molecular formula $C_{14}H_{14}N_3O_3SNa$ and molecular weight of 327.33(g/mol) (Gainiet al., 2008).

The dissociation constant, pKa, of MO is 3.47 in water at 25 °C; hence, MO exists predominantly as monovalent anions at pH values above 3.47 (Puthoor, 2017). MO has a molecular size of 1.19 nm × 0.68 nm × 0.37 nm and is non-volatile like all sulfonated azo dyes. All acid-base indicator occurs in two different tautomeric forms which are the benzenoid form and the quinonoid form; each of these forms has different colors and structures. For MO, the quinonoid form is red, present predominantly at pH above

3.1 and the benzenoid form is yellow, present predominantly at pH above 4.4 (Puthoor, 2017). The entire color change of MO occurs in acidic conditions between the pH ranges of 3.1 to 4.4 (Puthoor, 2017). The structures of MO in an acidic and basic medium are shown in Figure 2.1 below. On adding OH^- ions, a hydrogen ion will be lost from the N-N bond between the benzene rings (Puthoor, 2017).

On inadvertently entering the body through ingestion, Methyl Orange is metabolized by intestinal microorganisms into aromatic amines. Reductive enzymes in the liver can also catalyze the reductive cleavage of the azo linkage to produce aromatic amines and can even lead to intestinal cancer (Gainiet al., 2008).

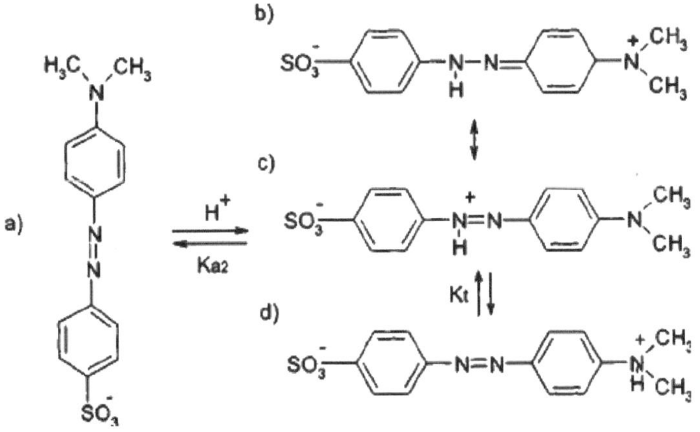

Figure 2. 1. Methyl Orange in its (a) Alkaline form and (b) monoprotonated zwitterionic structures

2.3 Adsorption process

Adsorption process occurs when a gas or liquid solute cleaves to a surface (adsorbent) forming a molecular or atomic film (adsorbate). This process differs totally from absorption which is about diffusing of a substance into a liquid or solid to form a solution.

Adsorption has been found to be superior to other conventional techniques to purify drinking water or clean up hazardous waste in wastewater, because of its initial cost, flexibility and simplicity of design, ease of operation and insensitivity to toxic pollutants (Uddin et al.; 2009). Adsorption is one of the applications of surface chemistry which resulted from surface energy i.e. atoms on the surface of adsorbent are attracted to adsorbates.

2.3.1 Types of adsorption

Adsorption can be classified into two types, which are physico-sorption and chemisorption. The differences between them are highlighted below:

Table 2. 2. Difference between physico-sorption and chemisorptions

Physico-sorption	Chemisorptions
Involved Van der Waals force between adsorbate and adsorbent	Involves formation of chemical bonds between adsorbate and adsorbent
Multi-molecular layer is formed	Monolayer is formed
Low enthalpy of adsorption	High enthalpy of adsorption
Reversible	Irreversible
No activation energy is involved	It involves activation energy
Equilibrium is achieved quickly	It takes longer time to achieve equilibrium
It occurs in solid/gas system	It occurs in highly specific process
Changes in the electronic states of adsorbent and adsorbate is minimal	Changes in the electronic states of adsorbent and adsorbate can be detected by suitable physical means
Typical binding energy is about 10-100meV	Typical binding energy is 1-10eV

Source: Butt *et al.*, (2003); Myers, (1999)

2.3.2 Alternative Adsorbents

This is the use of materials, other than the conventional adsorbents like activated carbon, which contain various binding capacities to accumulate dyes and other adsorbates. Bio-sorbents fall into this category as they have been shown to be ideal alternatives for decontamination of dyecontaining effluents.

Adsorption using alternative adsorbents typically involves a solid phase (biosorbent; adsorbent; biological material) and a liquid phase (solvent) containing a dissolved species to be sorbed (adsorbate, metals/dyes). Due to the high affinity of the adsorbent for the adsorbate species, the adsorbate is attracted to the adsorbent which are used to sorb heavy metals/dyes e.g. industrial by-product, agricultural wastes, and micro-organism and so on.

The use of alternative adsorbents has advantages compared with convectional techniques (Volesky, 1999). It is therefore listed below:

- **Cheap:** The adsorbent's cost is low since they are made from abundant or waste material.
- **Metal/Dye selective:** The affinity of the adsorbent for the adsorbate species can be more or less selective on different dyes. The dyes selectivity depends on type of adsorbent, mixture in the solution, type of adsorbent preparation and physical or chemical treatment.
- **No sludge generation**: There is a secondary problem such as sludge generation which is associated with the convectional techniques such as chemical precipitation. But in the use of alternative adsorption process, there is no such problem.
- **Regenerative:** These adsorbents can be reused, after the dye is removed.
- **Dye recovery:** It is highly possible to recover dye after they were sorbed from the solution.
- **Competitive performance:** Bio-sorption is capable of a performance comparable to the most similar technique; ion exchange treatment is rather costly making low cost of bio-sorption a major factor.

2.3.3. Factors affecting adsorption

Bio-sorption process is affected by effect of pH, particle size, contact time, temperature, adsorbent dosage, metal concentration, etc.

2.3.3.1 Effect of pH

This parameter has affected adsorption process in such a way that at low pH, concentration of proton is high and the ion exchange sites become solidly protonated, but when the pH solution increases, there is an increase in cation biosorption (Himanshu Agarwal *et al.*, 2010).

2.3.3.2 Effect of Contact Time

Contact time affects adsorption process in a way that increase in contact time will leads to increase the amount adsorbed ions. As the equilibrium reached, increase in contact time will not show any increase in adsorption (Himanshu Agarwal *et al.*, 2010).

2.3.3.3 Effect of temperature

Temperature affects adsorption process in a way that at 20-30^{0}C, the dye uptake increases within this temperature range and decreases with increase in temperature above a critical value. Increase in temperature affects the ions kinetic energy in manner that it will increase it and makes it easier for the metal ions to be attached to the adsorbent surface (Himanshu Agarwal *et al.*, 2010).

2.3.3.4 Effect of adsorbent dosage

This parameter affects the adsorption process in a way that increase in the adsorbent dosage will leads to decrease in dye uptake. This is because, the increase in the surface area of the adsorbent leads to increase in the number of sites present for binding (Himanshu Agarwal *et al.*, 2010).

2.3.3.5 Effect of dye concentration

This parameter is one of the factors that affect adsorption process in a way that as the dye concentration increases the amount of adsorbed ions decreases. This is because, there is greater driving force at higher concentration between the solid and liquid interface (Himanshu Agarwal *et al.*, 2010).

2.3.4 Agricultural by-products as Adsorbents

Agricultural by-products composed of cellulose, hemicelluloses, lignin, proteins, simple sugar, water, starch, hydrocarbons, lipids and various functional groups.

The presence of cellulose, hemicellulose, pectins and lignin in the cell- wall of the agricultural by-products, provided different functional groups such as hydroxyl, sulphonic, sulphydryl, amido, carboxylic, carbonyl, amino and nitro groups for adsorbate sorption (Qaiser *et al*., 2007).

Some of them include; sawdust (Ajmal *et al.,* 1996; Dakiky *et al.,* 2002;), coir pith (Parab, 2006; Unnithan*et al.,* 2004), straw (Chun *et al.,* 2004), wheat bran (Dupont and Guilon, 2003), cork powder (Machado *et al.,* 2002), sugar beet pulp (Altundogan, 2005), rice bran (Oliveira *et al.,* 2005), rice hulls (Marshal and Wartelle, 2004), bark (Sarin and Pant, 2006), wheat bran (Farooq *et al.,* 2010) and nut shells (Nasemajad *et al.,* 2005).

2.3.5 Industrial by-products as Adsorbents

Widespread industrial activities generate a lot of solid waste materials as by-products. Some of these materials are being utilized while others find no proper utilization and are being dumped elsewhere. Industrial waste is available almost free of cost and causes a major disposal problem. If industrial solid waste can be used as a low-cost adsorbent, it will provide a double fold advantage to environmental pollution. Firstly, the volume of waste material could be reduced and secondly the low-cost adsorbent if developed can reduce the pollution of sewage at a reasonable cost and in the view of low-cost adsorbent it will not be necessary to regenerate the spent materials (Bhatnagar and Sillanpaa, 2010).

2.3.5.1 Fly Ash

Fly ash is derived from the combustion of coal in thermal power plants. The major components of fly ash are alumina, silica, iron oxide, calcium oxide, magnesium oxide and residual carbon. One of the main advantages of fly ash over the other adsorbents is that it is in abundance and easily available to make it a strong choice in the investigation of an economical way of dye removal. Another advantage is that it could easily be solidified after the pollutants are adsorbed because it contains pozzolanic particles that react with lime in the presence of water to form cementation calcium-silicate hydrates (Singh, 2006, Sivamaniet *et al.*, 2009). Because of their low cost and local availability, industrial solid wastes such as fly ash, metal hydroxide sludge, fly ash and red mud are classified as low-cost materials and can be used as adsorbents for dye removal (Nanasivayam and Sumithra, 2005).

2.3.6 Inorganic Materials as Adsorbents

Metal oxides nanoparticles, clays, ceramics and minerals are used as adsorbents in the sequestration of dyes from its aqueous solution.

Clays are natural aluminosilicate having a few metal ions and organic compounds. Clays are available as soils sediments, rocks, and water. The use of clays is considered to be good because of its large surface area; high cationic exchange capacity, chemical and mechanical stability and layered structure (Aries *et al.*, 2005). Also, they are available abundantly at a lower cost. Natural clays are usually used for the removal of cationic dyes such as methylene blue due to their natural negative charged; however modifications to the surface of clay using surfactants can change the surface charge of clay from negative to positive (Erraiset *et al.*, 2012). These modifications enhance the adsorption of anionic dyes. Various types of clays have been studied by researchers in the removal of textile dyes and metal ions such as Reactive Red 120 by raw clay (Erraiset *et al.*, 2012), Brilliant Green dye by red clay (Rehmanet *et al*, 2013), Congo red by sodium bentonite, kaolin and zeolite (Vimonseset *et al.*, 2009), zinc ions by kaolin (Aries *et al.*, 2009) and Methylene Blue by montmorillonite clay (Almeida *et al.*, 2009).

2.3.7 Adsorption of Methyl Orange on various Adsorbents

Mohammadi *et al.* (2011) studied the adsorption behavior of MO onto mesoporous carbon material with a BET surface area of 940m^2/g and a pore volume of 0.96cm^3/g. The adsorption capacity was 331mg/g at a pH of 3. This study also reported an increase in adsorption capacity with a decrease in pH due to increase in positive charges on the surface favoring the adsorption of anionic MO. Chen *et al.* (2010) also reported a reduction in the adsorption capacity of a carbon material for MO when the pH was increased from 4.9 to 5.8, whereas the adsorption capacity remained constant outside this range. Xu *et al.* (2010) reported an enhanced photocatalytic degradation of MO when nanocomposites of TiO_2 impregnated on multi walled carbon nanotubes were used.

Yao *et al.* (2011) demonstrated that the maximum adsorption of MO was obtained at a pH of 2.3 and attributed the higher removal of MO to the electrostatic interactions between the anionic MO and positively charged carbon surface. The decrease in adsorption capacity with increasing pH was caused due to the decrease in positive charges on the carbon surface and the increase in competition between the anionic MO and OH^- ions in the solution. Adsorption of MO, arsenite, and arsenate was increased by creating positive charges or electron acceptor moieties on adsorbents obtained from 14 maize wastes by nitrogen containing substances in the research work of Elizalde-González *et al.* (2008). However, the results of Yu *et al.* (2012) contradict these observations in which they observed that the adsorption of MO was not sensitive to the pH of the solution and remained constant in the pH range of 3-10 irrespective of the change in the surface charges of the carbon, suggesting that electrostatic interaction was not helpful for the adsorption of MO.

2.3.8. Modification of adsorbents

This is the process of improving adsorption capacity (active site) and reduces the amount of soluble organic compounds released to the environment. The modification of the adsorbent can be done by physical and chemical modification methods.

2.3.8.1. Physical modification method

It is the modification process of the adsorbent, which is often referred to as pre-treatment without using chemicals. This can be achieved by grinding, carbonizing, high energy radiation and steam exploding etc.

2.3.8.2. Chemical modification method

It is the process of increasing adsorption capacity of the adsorbent by modifying the adsorbent with chemicals. This process is effective because they added more of active sites to the adsorbent. Most of the chemicals added are mineral and organic acid, base, organic compounds and oxidant. This can be achieved by esterification, etherification, halogenations, oxidation, and graft copolymerization.

Absorbents can be characterized by using spectroscopic techniques, which deals with interaction of energy with the adsorbent to perform an analysis. These spectroscopic instruments are; Fourier Transform Infrared (FTIR), Scanning Electron Microscopy - Energy dispersive Spectroscopy (SEM-EDS), XAFS (X-ray Absorption Fine Structure).

2.4 Adsorption equilibrium models (Adsorption isotherm)

This is the graphical illustration or representation of the relationship between the amount of gas or liquid molecules adsorbed by a given adsorbent and the pressure in case of gas or concentration in case of liquid. Adsorption isotherms focus mainly on systems where the adsorbate particles are mostly concentrated on the surface of an adsorbent. The Langmuir isotherm model describes the dependence of the surface coverage of an adsorbed species on the pressure/concentration of the species at a fixed temperature. The Freundlich isotherm model describes physical adsorption in solution while the BET isotherm applies to multi-layer adsorption.

2.4.1 Langmuir isotherm model

Langmuir isotherm model predicts the existence of monolayer coverage of the adsorbate on the outer surface of adsorbent. The Langmuir isotherm equation is given by:

Ce = 1 + 1 Ce .... (1)

qe qmK$_L$ qm

Where C_e, q_e, q_m and K_L is the equilibrium concentration of adsorbate (mg/L), amount of adsorbate adsorbed at equilibrium (mg/g), monolayer adsorption capacity of the adsorbent (mg/g) and the Langmuir adsorption constant (L/mg) respectively.

Langmuir adsorption isotherm is the graphical plot of $\mathbf{C_e/q_e}$ against $\mathbf{C_e}$ which gives a straight line with slope of $\mathbf{1/q_m}$ and intercept of $\mathbf{1/(K_L q_m)}$. It works for chemisorptions in both gas-solid and liquid-solid interphase.

The Langmuir parameter called dimensionless separation factor R_L can be used to predict the affinity of the adsorbent (crushed ceramics) towards the adsorbate (methyl orange). R_L is given by:

RL= 1/ (1 + K$_L$Co) .... (2)

The dimensionless separation factor R_L can be used to interpret the isotherm shape.

Table 2. 3. The shapes of Langmuir isotherm

Value of R_L	Type of adsorption
$R_L>1$	Unfavourable
$R_L=1$	Linear
$R_L=0$	Irreversible

(Amuda *et al.*, 2007)

2.4.2 Freundlich isotherm model

In, 1906, Freundlich assumes a heterogenous surface (multi-layer) with a non-uniform distribution of heat of biosorption over the surface. This model is for a single solute system in which it is expressed as:

$qe = KFCe^{1/n}$ ....(3)

It can be simplified further as:

$logqe = 1/nlogCe + logK_F$ ....(4)

Where q_e, C_e, 1/n and K_F is the amount of adsorbate adsorbed per unit mass of adsorbent (mg/g), equilibrium concentration of the adsorbate (mg/L), adsorption intensity and Freundlich constant ((mg/g $(L/mg)^{1/n}$) respectively.

Freundlich isotherm model is the graphical plot of **$\log q_e$** versus **$\log C_e$** which gives a slope of **1/n** and intercept of **$\log K_F$**.

To determine whether the data will fit in for Freundlich isotherm model, it is determined as follows;

Table 2. 4. Freundlich slope range and its adsorption types

Slope range	**Type of adsorption**
$0<n<1$	Favourable
1/n is greater than or equal to 0	More heterogeneous
1/n is less than or equal to 1	Unfavourable

2.4.3. Brunaer-Emmett-Teller (BET)

In 1938, BET assumed that the adsorbate molecules could be adsorbed in more than one layer thickness on the surface of adsorbent. It is expressed as:

qe= KB Ceqm/ (Cs-Ce) [1 + (Kb -1) (Ce/Cs)] ……………………………………..(5)

Where q_e, q_m, K_b, C_s and C_e is the amount of adsorbate adsorbed at equilibrium (mg/g), maximum adsorption capacity (mg/g), BET constant, solute concentration at the saturation of all layers (mg/L) and equilibrium concentration of adsorbate (mg/L) respectively. It can be simplified further as:

Ce/ [(Cs-Ce) qe] = 1/ (Kb qm) + (Kb – 1 / Kbqm) (Ce/Cs) …………………….(6)

BET isotherm model is the graphical plot of $\mathbf{C_e/ [(C_s\text{-}C_e)q_e]}$ versus $\mathbf{C_e/C_s}$ which gives a straight line with slope of $\mathbf{(K_b – 1/K_bq_m)}$ and intercept of $\mathbf{1 / K_bq_m}$.

2.5 Adsorption kinetic studies

Kinetic study gives information on the reaction pathway and the mechanism of the reaction. Adsorption kinetic study relates the relationship between concentration of adsorbate in solution and adsorption rate.

2.5.1 Pseudo-first order kinetic model

This model assumes that the rate of change of the adsorption of solute is proportional to the difference in saturation concentration and the amount of solid uptake. It is expressed as:

dq/dt = k1 (qe– qt) ………………………..…………………………………...(1)

By integration, it can be simplified further as:

ln (qe – qt) = lnqe – k1t …………………………………………………………...(2)

Where q_t and qe are the masses of adsorbate adsorbed at any time t and at equilibrium in (mg/g) respectively, K_1 is the rate constant (min^{-1}).

Pseudo-first order kinetic model is the graphical plot of **ln (q_e-q_t)** versus **t** which gives the straight line with slope of **k_1** and intercept of **lnq_e**.

2.5.2 Pseudo-second order kinetic model

The pseudo-second order kinetic model is based on the adsorption capacity onto a solid phase. It is expressed as:

$$dqt/dt = k_2(qe-qt)^2 \quad \text{.....(3)}$$

By integration, it can be simplified further as:

$$1 / (qe - qt) = 1/qe + k_2 \quad \text{.....(4)}$$

By re-arrangement;

$$t/qt = 1/h + (1/qe)t \quad \text{.....(5)}$$

$$h = k_2qe^2 \quad \text{.....(6)}$$

Where h is the initial sorption rate (mg/g min) and K_2 is pseudo second-order rate constant (g/mg min).Pseudo-second order is the graphical plot of t/qt against t which gives a straight line with $1/q_e$ as slope and 1/h as an intercept.

CHAPTER THREE.

MATERIALS AND METHODS

This section details the equipment, reagents and the methods used in the study of the adsorption of methyl orange on crushed ceramics adsorbents.

3.1 Materials

3.1.1 Apparatus and equipment

Apparatus used are;

- 250-mL beakers
- 250-mL conical flasks
- 1000-mL standard flasks
- spatula
- weighing papers
- sieve
- filter papers
- pipette

Equipment used are;

- Analytical balance readability of 0.0001g
- UV-Visible Spectrophotometer
- Centrifuge
- Orbital shaker
- Refrigerator
- Shaking water bath
- pH meter (pH-107)

3.1.2 Chemical reagents

Chemical reagents used are;

Methyl orange dye

Sodium hydroxide

Hydrochloric acid

Distilled water

3.2 Methods

3.2.1 Adsorbent preparation

Ceramics was purchased from Oja-oba market in Akure, crushed with grinding stone and sun-dried for three days. The dried crushed ceramics was then sieved with 63μm laboratory test sieve to produce homogenized particle size and stored in an air-tight container until usage.

Figure 3. 1. Crushed Ceramic Adsorbent

3.2.2 Preparation of reagents

3.2.2.1 Preparation of Methyl Orange Dye Stock Solution

100mg/L of Methyl orange dye solution was prepared by adding 100mL of the standard solution of methyl orange dye stock in 1000mL standard flask and added up to the mark with distilled water.

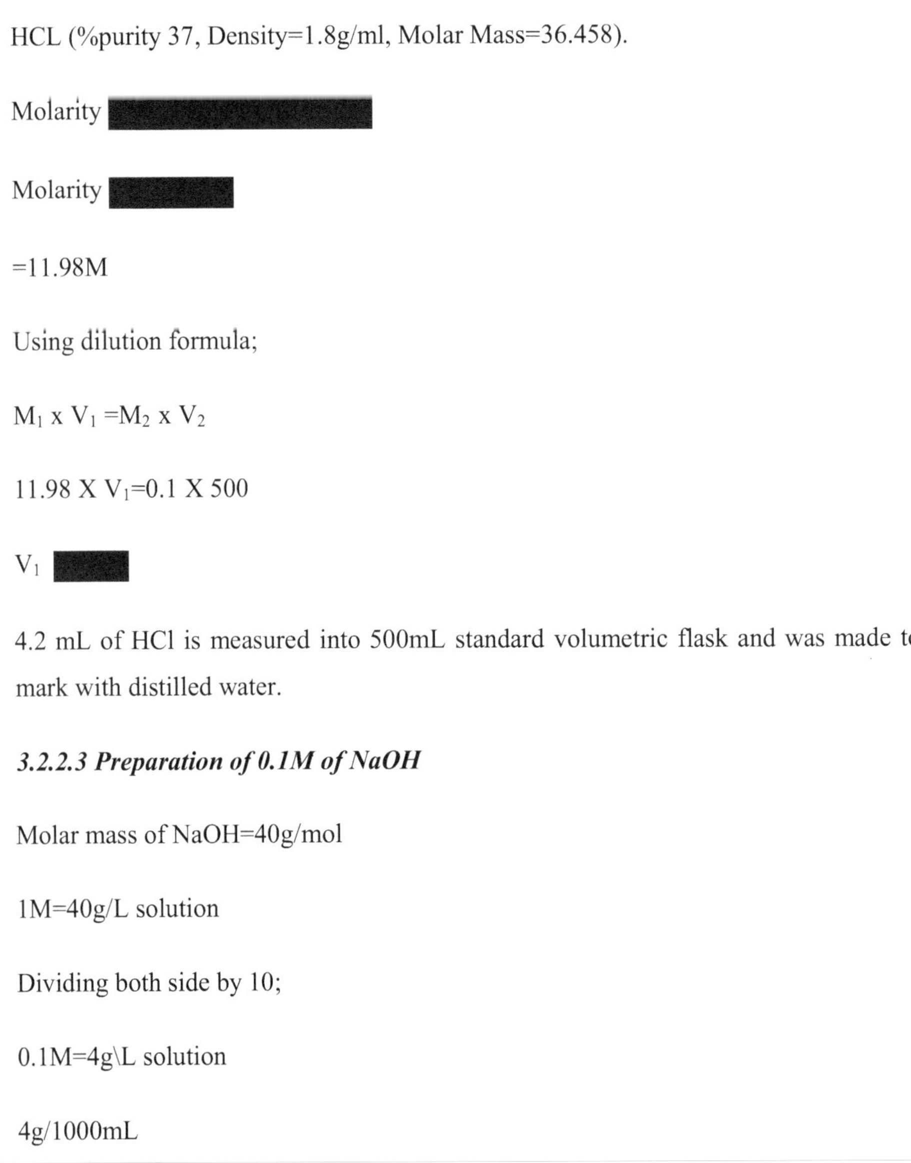

3.2.2.2. Preparation of 0.1M HCL

HCL (%purity 37, Density=1.8g/ml, Molar Mass=36.458).

Molarity

Molarity

=11.98M

Using dilution formula;

$M_1 \times V_1 = M_2 \times V_2$

$11.98 \times V_1 = 0.1 \times 500$

V_1

4.2 mL of HCl is measured into 500mL standard volumetric flask and was made to mark with distilled water.

3.2.2.3 Preparation of 0.1M of NaOH

Molar mass of NaOH=40g/mol

1M=40g/L solution

Dividing both side by 10;

0.1M=4g\L solution

4g/1000mL

2g/500mL

2g of NaOH is measured into 500mL standard flask, and then distilled water was added to make it up to 500mL mark of standard flask.

3.2.2.4 Preparation of working standard solutions for UV-Visible Spectrophotometer

Ten working solutions were prepared from the stock solution using the dilution formula:

$$C_1V_1 = C_2V_2$$

Where C_1 is the concentration of the standard solution (100 mg/L), V_1 is the required volume of

the standard, C_2 is the required working standard concentrations (10mg/L to 100mg/L) and V_2 is 10mL.

3.2.3 Effect of pH

To test for the effect pH on the adsorption process, 1.0g of the adsorbent was weighed into 7 pre labeled beakers, and 25mL of 100mg/L of methyl orange dye solution were added into each beaker. Thereafter, the pH of the mixtures was adjusted to 1.0, 2.0, 3.0, 4.5, 6.0, 7.0, and 8.0 for the seven beakers respectively, using 0.1M HCl and 0.1M NaOH solutions. After that, the samples were agitated using orbital shaker at 100rpm for 15 minutes. The samples were withdrawn from the shaker after the specified period of time, centrifuged and the dye solution was separated from the adsorbent by decantation. The absorbance of supernatant solution was measured using UV-Vis Spectrophotometer at wavelength 464 nm (Lijuan wu and Xuewen Liu 2021).

3.2.4 Effect of contact time

The beakers to be used were labeled and 1.0g of the adsorbent was weighed into 10 beakers, after which 25mL of 100mg/L methyl orange dye solution were added into each beaker. After this, each sample was adjusted to the optimum pH, and the samples were shaken on the orbital shaker at different time interval. The time intervals used, in minutes, were: 2, 5, 10, 15, 20, 25, 30, 60, 90, and 120. Each sample was withdrawn at each time interval, centrifuged and the dye solution was separated from the adsorbent by decantation. The absorbance of supernatant solution was measured using UV-Vis Spectrophotometer at wavelength 464 nm (Lijuan wu and Xuewen Liu 2021).

3.2.5 Effect of Adsorbent dosage

The beakers were labeled and different adsorbent dosages (0.1, 0.5, 1.0, 1.5, 2.0, and 2.5g) were weighed into them respectively. Thereafter, 25mL of 100mg/L Methyl orange dye solution were added to the each sample in the beaker. After this, each sample was adjusted to the optimum pH and agitated. The samples were withdrawn from the shaker after the optimum contact time, centrifuged and the dye solution was separated from the adsorbent by decantation. The absorbance of supernatant solution was measured using UV-Vis Spectrophotometer at wavelength 464 nm (Lijuan wu and Xuewen Liu 2021).

3.2.6 Effect of the initial concentration

1.0g of the crushed ceramic adsorbent was weighed into the labeled beakers. Thereafter, 20mL of different initial dye concentrations of 50, 100, 150, 200, 250 and 300 mg/L were pipetted into pre-labeled beakers. After this, the samples were adjusted to the optimum pH and agitated. The samples were withdrawn from the shaker after the optimum contact time, centrifuged and the dye solution was separated from the adsorbent by decantation. The absorbance of supernatant solution was measured using UV-Vis Spectrophotometer at wavelength 464 nm (Lijuan wu and Xuewen Liu 2021).

3.2.7 Effect of Temperature

The crushed ceramic adsorbent 1.0gwas weighed into the labeled beakers. Thereafter, 20mL of 50mg/L of methyl orange dye solution were added respectively into each beaker at room temperature, 35, 45, 55 and 65°C by batch experiments, at constant pH and were shaken on the water bath shaker for the same contact time. The supernatant was collected after centrifugation and decantation; and measured using UV-Vis Spectrophotometer at wavelength 464 nm (Lijuan wu and Xuewen Liu 2021). This procedure is repeated for 100, 150, 200, 250 and 300 mg/L of dye solution respectively for the elucidation of the kinetic characteristics of the adsorption.

3.3.8 Statistical Analysis (Data Analysis)

All the experiments were conducted in duplicates and the data were presented as mean values ± standard deviation of the duplicate determinations, and as graphs and curves using Microsoft

Excel, 2016.

CHAPTER FOUR.

RESULTS AND DISCUSSION

4.1 CHARACTERIZATION OF THE ADSORBENT

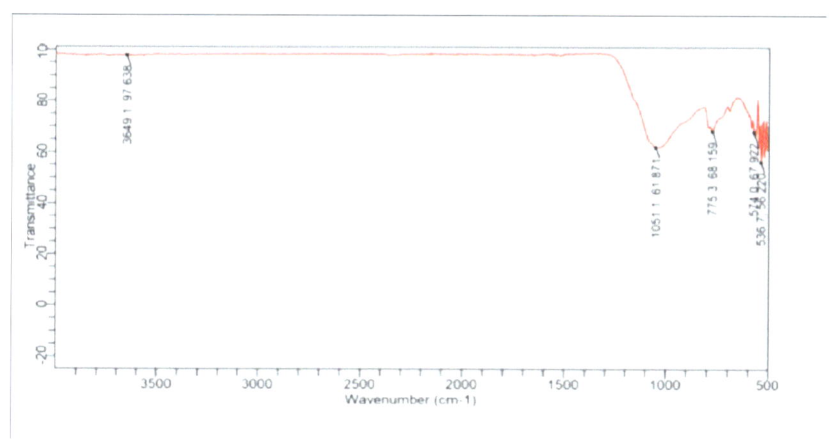

Figure 4. 1. FTIR spectra of adsorbent before adsorption

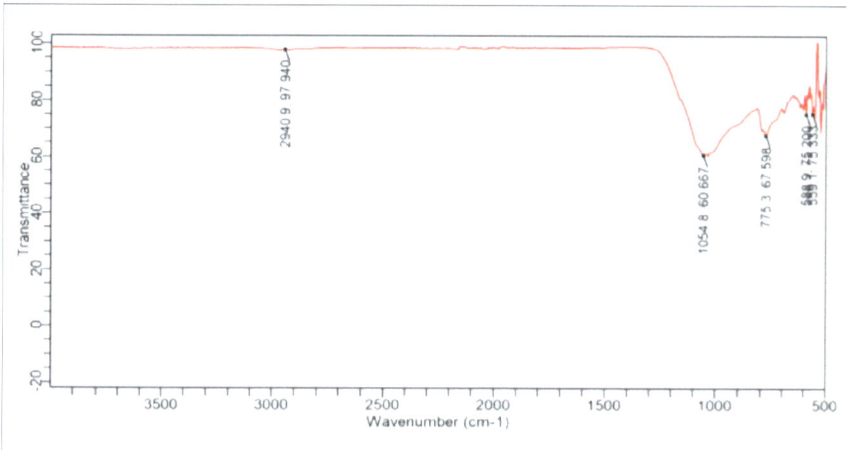

Figure 4. 2. FTIR spectra of adsorbent after adsorption

Figures 4.1 and 4.2 show the FTIR spectra of the adsorbent, before and after adsorption. The two spectra show similar peaks, with the second spectra having a slight shift to the left. The peak at 3649.1 cm^{-1} in the before-spectra was due to the stretching of O-H group as a result of intermolecular and intra-molecular hydrogen bonding of polymeric compounds such as alcohols or pectin, hemicelluloses, cellulose and lignin (Iqbal *et al.*, 2008). The peak around 2940.9 cm^{-1} in the after-spectra corresponds to the stretching vibrations of C-H bond of methyl (CH_3), methylene ($=CH_2$), methoxy groups (OCH_3) (Feng *et al.*, 2008). The intense peak at 1051.1cm^{-1} corresponded to the C-O stretching of alcohol or carboxylic acid (Ngah and Hanafiah, 2007).

The peak at 775 cm^{-1} corresponds to the C-H stretching in aromatic ring (Khan *et al.*, 2011).

4.2 OPTIMIZATION OF FACTORS AFFECTING THE ADSORPTION OF METHYL ORANGE ON CRUSHED CERAMICS

4.2.1 Effect of pH

The effect of pH on removal of methyl orange in a solution by crushed ceramics adsorbent was studied at pH of 2 to 8. Figure 4.3 below shows the removal efficiency (%) with their corresponding pH values.

The removal efficiency (%) of methyl orange by the adsorbent was generally higher at the lower pHs of 1 (98.29%) and 2 (99.31%) compared to the other pHs. This may be due to the fact that at lower pH, the surface of the adsorbent becomes more protonated, having more H^+ from the 0.1M HCl and other positive ions that may be there as impurities. This in turn makes methy orange dye, which is an anionic dye, bind with the surface of the adsorbent easily (Sathya *et al.*, 2017). With increase in pH, the hydroxyl ion (OH^-) from the 0.1M NaOH competes with the anionic dye for sorption sites, thereby reducing its adsorption efficiency (Sathya *et al.*, 2017).

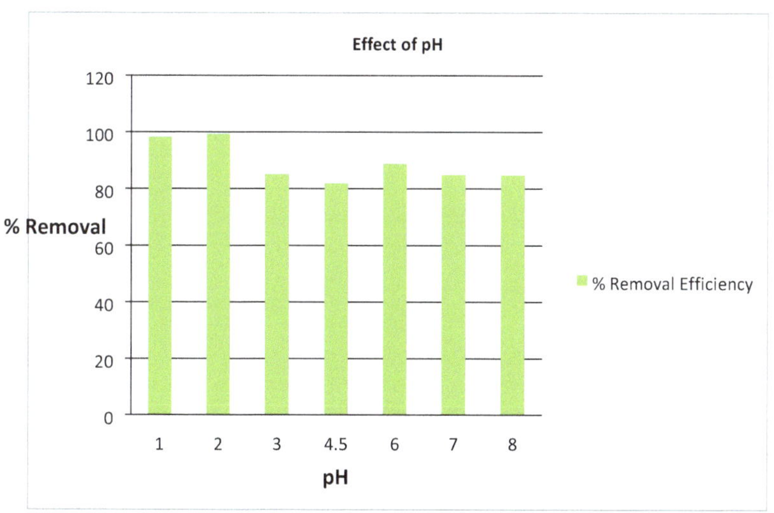

Figure 4. 3. Effect of pH on removal of methyl orange with crushed ceramics.

Initial concentration = 100mg/L, Agitation speed = 100rpm, Contact time = 2hours, Adsorbent dosage = 1.0g.

The optimum pH for the rest of the experiments was thus selected as pH 2 with removal efficiency of 99.31% and was chosen as one of the parameters in the next optimization factors.

4.2.2 Effect of contact time

Contact time plays a key role in the adsorption process and it often determines the amount of dyes adsorbed on an adsorbent. It is an important parameter for successful use of the adsorbents for practical applications. Rapid adsorption is among the desirable parameters in choice of a suitable adsorbent. The effect of contact time on adsorption of methyl orange dye was investigated and the results are displayed in figure 4.3 below.

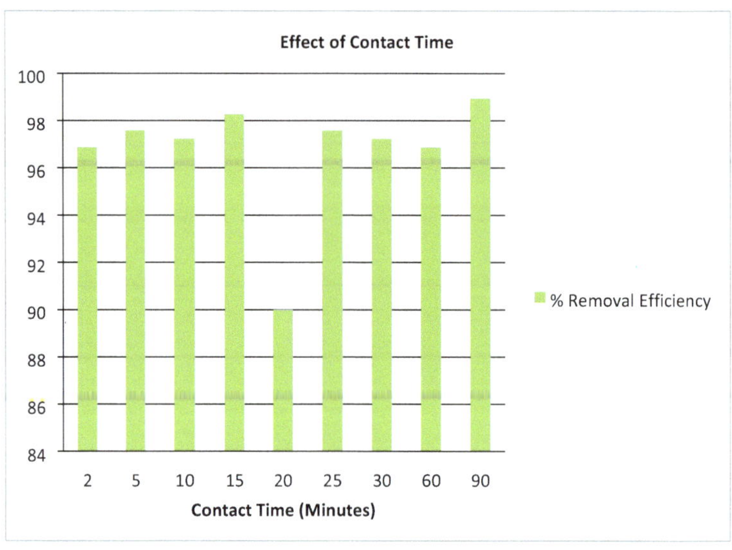

Figure 4. 4. Effect of contact time on removal of methyl orange dye

pH = 2.0, Initial concentration = 100mg/L, Agitation speed = 100rpm, Adsorbent dosage = 1.0 g.

As shown in the figure 4.3, the rate of sorption of methyl orange dye generally increases with increasing the contact time. At 90 minutes, the equilibrium is achieved as a result of the binding sites that became exhausted. The percentage removal gradually slowed down due to decreasing availability of active sites by Methyl orange (Ong Pick Sheen, 2011).

4.2.3 Effect of adsorbent dosages

The effect of adsorbent dosages ranged from 0.1g to 2.5g on the adsorption of Methyl orange are presented in figure 4.5.

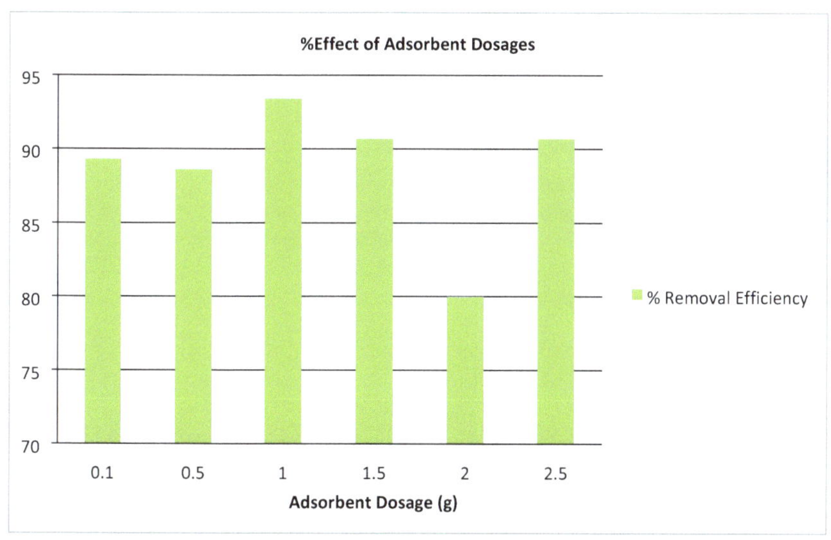

Figure 4. 5. Effect of the adsorbent dosage on removal of methyl orange dye.

pH = 2.0, Initial concentration = 100 mg/L, Agitation speed = 100rpm, Contact time = 90 Minutes

As it can be seen from Figure 4.5, adsorption of the Methyl orange dye increased from 89% to 94% with increase in adsorbent dosage from 0.1 to 1g. At adsorbent dosage of 2g, the percentage removal dropped which was as a result of the overlapping of the adsorption sites.

The results show that as the adsorbent dosage increases the affinity for methyl orange to bind with the binding sites decreased which was observed to be due to the overlapping of the negatively charged ions at the surface of the adsorbent (Gebrehawaria, 2016). An increase in the adsorbent dosage later increased with the percentage removal and reached its equilibrium again at 2.5g with % removal 91%, which was as result of the binding sites available have become exhausted.

4.2.4 Effect of the initial concentration

The effect of the initial concentration ranges from 50, 100, 150, 200, 250 and 300mg/L on the amount of methyl orange adsorbed is presented in figure 4.6.

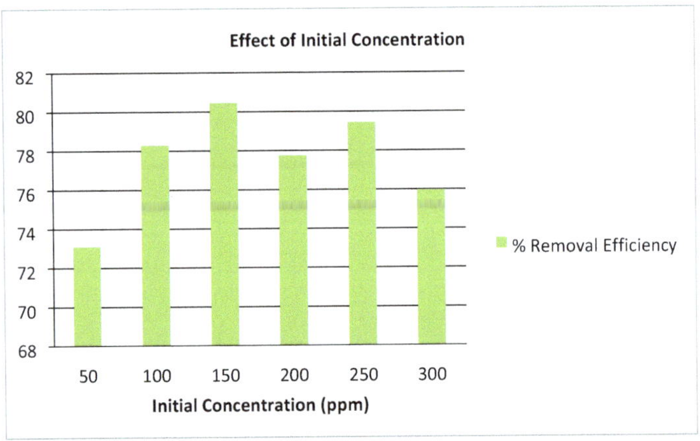

Figure 4. 6. Effect of the initial concentration on removal of Methyl orange.

pH = 2.0, Agitation speed = 100rpm, Contact time = 90 Minutes, Adsorbent dosage = 1.0g

It was observed in figure 4.6 that with increase in the initial Congo red concentration, the percentage removal first increased then decreased. The percentage removal decreased because at higher concentration the available sites on the adsorbent become limited. The adsorption capacity increased as dye concentration increase as a result of utilizing all available adsorption sites at higher concentration (Khan *et al.*, 2011).

4.2.5 Effect of Temperature

Figure 4.7 shows the effect of temperature on the adsorption efficiency.

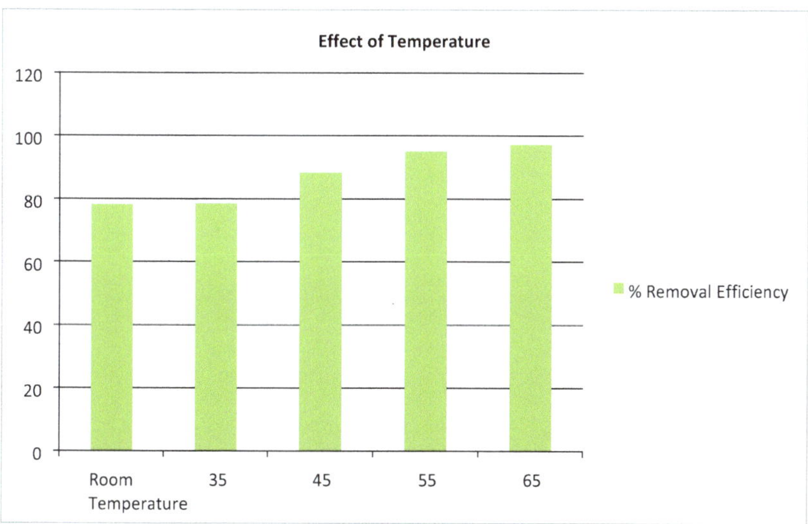

Figure 4. 7. Effect of the temperature on removal of methyl orange dyes, Eriochrome Black T and their binary mixture using untreated pulverized elephant grass.

pH = 2.0, Agitation speed = 100rpm, Contact time = 90 minutes, Initial concentration = 30mg/L, Adsorbent dosage = 1.0g.

4.3 ADSORPTION ISOTHERMS

4.3.1 Langmuir adsorption isotherm

The Langmuir isotherm model is a straight line graph which is a plot of C_e/q_e against C_e. For the sorption of methyl orange onto crushed ceramic adsorbent, the Langmuir isotherm model graph was presented in figure 4.8.

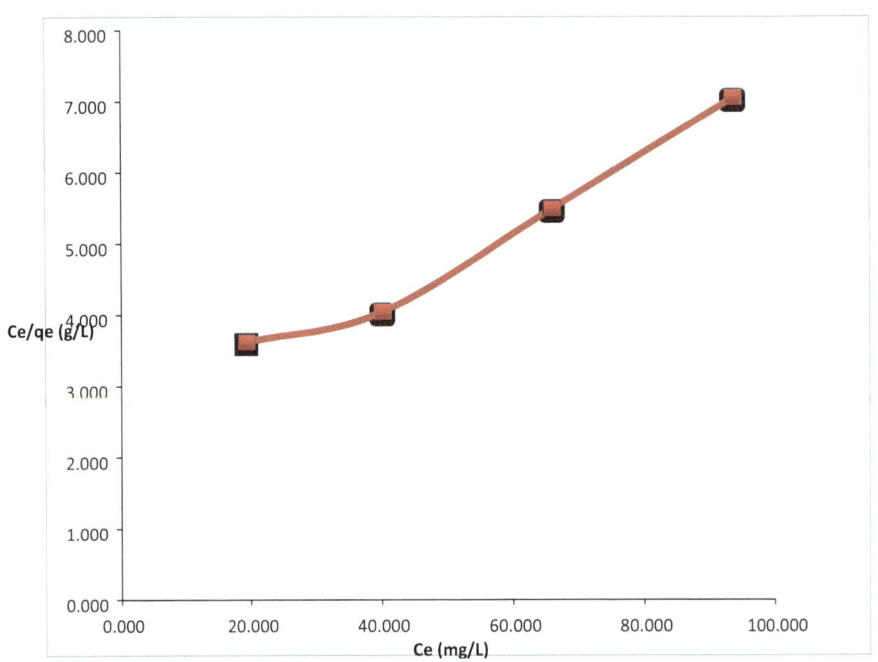

Figure 4. 8. Langmuir Isotherm for methyl orange adsorption

Table 4. 1. Langmuir isotherm parameters

Methyl Orange	
qmax	**K_L**
22.31	0.020

Table 4. 2. RL values for the adsorption of Methyl Orange mixture

Methyl Orange	
Initial Concentration	R_L
50	0.49
100	0.41
150	0.35
200	0.2

4.3.2 Freundlich adsorption isotherm

The Freundlich isotherm assumes a heterogeneous surface with a non-uniform distribution of heat of bio-sorption over the surface and a multilayer bio-sorption can be expressed (Freundlich, 1906). The Freundlich model was expressed as:

log qe = log K_F + 1/nlogCe..(1)

Freundlich isotherm model is a straight line graph and a plot of log q_e against log C_e. For the sorption of methyl orange onto crushed ceramic adsorbent, the Freundlich isotherm model graph was presented in figure 4.9.

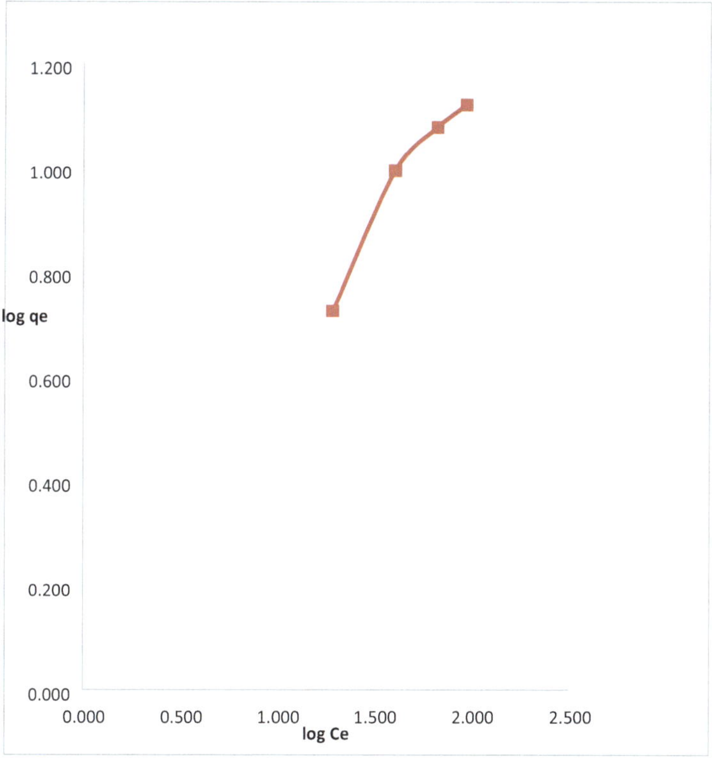

Figure 4. 9. Freundlich isotherm plot for the sorption of methyl orange onto crushed ceramics

Table 4. 3. Freundlich isotherm parameters

Methyl orange	
N	K_F
2.724	2.026

Results obtained show that the equilibrium data for the adsorption of methyl orange are best fitted for Freundlich isotherm because the N-value and their R^2 values are close to one. Chen *et al.*, (2010), however, proposed that the values of N in the range of 2 to 10 are good, 1 to 2 values are moderate and less than 1 are poor sorption characteristics. For the sorption of methyl orange, the multi-layer is formed which is the physic-sorption type of adsorption.

4.4 ADSORPTION KINETICS

Adsorption kinetics provides the information about the mechanism of adsorption which is important for efficiency of the adsorption process.

4.4.1 Pseudo-First Order Model

This is the plot of log (q_e-q_t) against t in which K_1 and q_e are calculated from the slope and intercept respectively from the plot. Figure 4.9 shows the plot of pseudo-first order model at concentration of 100mg/L and at varied time (10, 15, 20, 25 minutes).

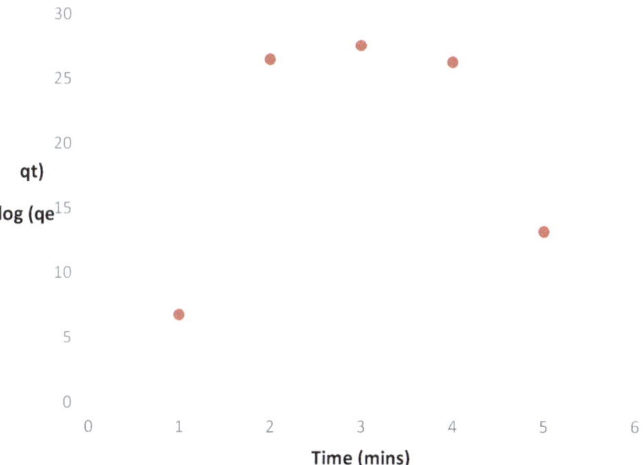

Figure 4. 10. Plot of pseudo-first order model of the adsorption of methyl orange onto modified mango leaves.

pH = 2.0, Initial concentration = 100mg/L, Agitation speed = 100rpm, Adsorbent dosage = 1.0g.

Table 4. 4. Pseudo-first order kinetic parameters for methyl orange

	Methyl orange	
q_e exp (mg/g)	**q_e cal (mg/g)**	**K1 (min-1)**
76.87	84.92	0.0074

4.4.2 Pseudo-Second Order Model

The pseudo-second order kinetic rate equation was expressed as (Ho & Mckay, 1998):

$dqt/dqt = k_2(qe - qt)^2$ ....(1)

$t/qt = 1/k_2qe2 + 1/qet$ ....(2)

This is the plot of t/q_t against t in which q_e and K_2 are calculated from the slope and intercept respectively from the plot. Figure 4.10 shows the plot of pseudo-second order model at concentration of 100mg/L and at varied time (10, 15, 20, 25 minutes).

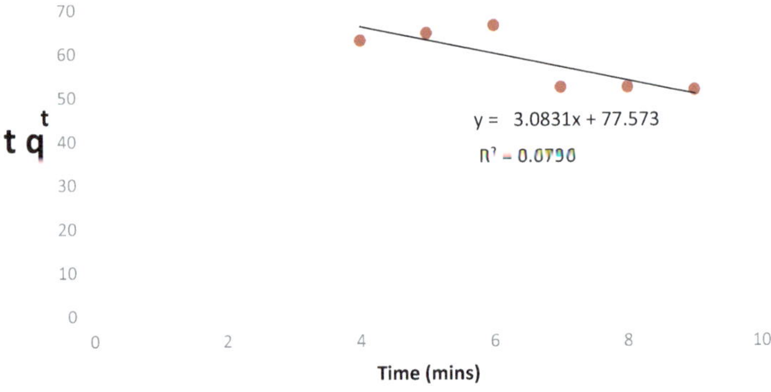

Figure 4. 11. Plot of pseudo second order model of the adsorption of methyl orange on crushed ceramic

pH = 2.0, Initial concentration = 100mg/L, Agitation speed = 100rpm.

Table 4. 5. Pseudo-second order kinetic parameters for the adsorption of mercury (II) ions, Congo red dye and their binary mixture by modified mango leaves

Methyl Orange		
q_e exp (mg/g)	q_e cal (mg/g)	K_2 (min^{-1})
-0.0017	45.455	71.45

By comparing Table 4.5 and 4.6, Pseudo-Second Order kinetics was seen as the best fitted for the adsorption of methyl orange onto crushed ceramics than Pseudo-First Order kinetics since the calculated equilibrium adsorption capacity value of methyl orange was close to the experimental equilibrium adsorption capacity value in Pseudo-Second Order kinetics than in Pseudo-First Order kinetics.

CHAPTER FIVE.

CONCLUSION

In this study, powdered ceramics was investigated for their performances in removing methyl orange from contaminated aqueous solutions. FTIR spectra of ceramics powder revealed that OH, C-H, C=C, C-O stretching were responsible for the adsorption. Effects of the different experimental parameters that influenced the efficiencies of the adsorbent have been evaluated and optimized.

Since wastewater from the textile and dye industries contain a lot of coloured organic compounds, this work was geared towards removing methyl orange dye at once from the contaminated aqueous solution. The investigation revealed that the adsorption capacity of the powdered ceramics on removal of high enough to be comparable to observed values in literature. Kinetic data obtained for the adsorption of methyl orange dye was fitted well with pseudo second order model.

5.1 RECOMMENDATION

To improve the adsorption capacity of the powdered ceramics on removal of dyes from the contaminated aqueous solutions; it is recommended that;

- the binding sites of the powdered ceramics be increased by treating it with other chemicals that could open up pore sites for more adsorption
- more studies to be conducted on the mechanisms of adsorption by use of Scanning Electron Microscopy (SEM) and Energy Dispersive Spectroscopy (EDS) in order to understand how the adsorption takes place.

REFERENCES

Ahmed, M.N. and Ram R.N. (1992). Removal of basic dye from wastewater using silica as adsorbent, *Environ. Pollut,* 77, 79–86.

Arami, M., Limaee, N.Y., Mahmoodi, N.M., and Tabrizi, N.S. (2006). Equilibrium and kinetics studies for the adsorption of direct and acid dyes from aqueous solution by soymeal hull. *J Hazard Mater*; 135(1-3): 171-9. doi: 10.1016/j. jhazmat.

Arshad, A., Iqbal, J., Ahmad, I., & Israr, M. (2018). Graphene/Fe3O4 nanocomposite: Interplay between photo-Fenton type reaction, and carbon purity for the removal of methyl orange. *Ceramics International*, *44*(3), 2643-2648.

Bhatti, H.N. and Nausheen, S. (2014). Equilibrium and kinetic modeling for the removal of Turquoise Blue PG dye from aqueous solution by low cost agro-waste. *Desalin Water Treat*; 55(7): 1934-44. doi: 10.1080/19443994.2014. 927799.

Chen, J., Shi, X., Zhan, Y., Qiu, X., Du, Y., & Deng, H. (2017). Construction of horizontal stratum landform-like composite foams and their methyl orange adsorption capacity. *Applied Surface Science*, *397*, 133-143.

Do M., Abak H. and Alkan M. (2009). Adsorption of methylene blue onto hazelnut shell: Kinetics, mechanism and activation parameters, *J. Hazard. Mater.,* 164, 172–181.

Do. J.S. and Chen. M.L. (1990). Decolourization of dye containing solutions by electrocoagulation, *J. of Appl.Electrochemistry*, 24, 785-790.

Gupta, V. and Suhas, K. (2009). Application of low-cost adsorbents for dye removal – A review. J Environ Manage; 90(8): 2313-42. doi: 10.1016/j.jenvman.2008.11.017.

Gupta.V.K and Rastogi.A (2007). 'Biosorption of lead from aqueous solutions by green algae *Spirogyra species*: Kinetic and equilibrium studies' *Journal of Hazardous Materials*, Vol.152, pp. 407-414.

Hosseini, S., Khan, M. A., Malekbala, M. R., Cheah, W., & Choong, T. S. (2011). Carbon coated monolith, a mesoporous material for the removal of methyl orange from aqueous phase: Adsorption and desorption studies. *Chemical engineering journal*, *171*(3), 1124-1131.

Iqbal M, Saeed A, Zafar SI (2009b). FTIR spectrophotometry, kinetics and adsorption isotherms modeling, ion exchange, and EDX analysis for understanding the mechanism of Cd2+ and Pb2+ removal by mango peel waste. *J. Hazard. Mater.*, 164: 161-171.

Jalil, A. A., Triwahyono, S., Adam, S. H., Rahim, N. D., Aziz, M. A. A., Hairom, N. H. H., ... & Mohamadiah, M. K. A. (2010). Adsorption of methyl orange from aqueous solution onto calcined Lapindo volcanic mud. *Journal of Hazardous Materials*, *181*(1-3), 755-762.

Khan, T.A., Rahman, R., Ali, I., Khan, E.A., and Mukhlif, A.A. (2014). Removal of malachite green from aqueous solution using waste pea shells as low-cost adsorbent-adsorption isotherms and dynamics. *Toxicol Environ Chem;*96(4): 569-78. doi: 10.1080/02772248.2014.969268.

Khan, T.A., Sharma, S., Khan, E.A., and Mukhlif, A.A. (2014). Removal of congo red and basic violet 1 by chir pine (Pinus roxburghii) sawdust, a saw mill waste: batch and column studies. *Toxicol Environ Chem*; 96(4): 555-68. doi: 10.1080/02772248.2014.959017.

Krysztafkiewicz, A., Binkowski, S. and Jesionowski T. (2002). Adsorption of dyes on a silica surface, *Appl. Surf. Sci.,* 199, 31–39.

Malik, R., Ramteke, D.S. and Wate, S.R. (2006). Physico-chemical and surface characterization of adsorbent prepared from groundnut shell by ZnCl2 activation and its ability to adsorb colour, *Indian Journal of Chemical Technology,* 13,319-328.

Meshko, V., Markovska, L., Mincheva M. and Rodrigues. (2001). Adsorption of basic dyes on granular activated carbon and natural zeolite, 35, 3357–3366.

Ni, Z. M., Xia, S. J., Wang, L. G., Xing, F. F., & Pan, G. X. (2007). Treatment of methyl orange by calcined layered double hydroxides in aqueous solution: adsorption property and kinetic studies. *Journal of Colloid and Interface Science*, *316*(2), 284-291.

Ofomaja, A.E. (2008). Kinetic study and sorption mechanism of methylene blue and methyl violet onto mansonia (Mansonia altissima) wood sawdust. *Chem Eng J;*143(1-3): 85- 95. doi: 10.1016/j.cej.

Ozdemir, O., Armagan, B., Turan, M., and Celik M.S. (2004). Comparison of the adsorption characteristics of azoreactive dyes on mezoporous minerals, *Dye. Pigment.,* 62, 49–60.

Subbaiah, M. V., & Kim, D. S. (2016). Adsorption of methyl orange from aqueous solution by aminated pumpkin seed powder: Kinetics, isotherms, and thermodynamic studies. *Ecotoxicology and environmental safety*, *128*, 109-117.

Ushakumary, E. R. (2013). "Waste Water Treatment Using Low Cost Natural Adsorbents". Division of Safety and Fire Engineering, Cochin University of Science and Engineering,

Kerala, India. A thesis submitted in partial fulfilment of the requirement for the award of Doctor of Philosophy”. pp 21-29.

Yao, C., Zeng, Y., Cao, Y., Li, W., & Zheng, X. (2010). Adsorption of methyl orange on polyaniline/attapulgite nanocomposites. *Guisuanyan Xuebao(Journal of the Chinese Ceramic Society)*, *38*(4), 671-677.

www.ingramcontent.com/pod-product-compliance
Ingram Content Group UK Ltd.
Pitfield, Milton Keynes, MK11 3LW, UK
UKHW061023310726
14090UKWH00023B/60

* 9 7 9 8 8 9 2 4 8 4 0 1 5 *